IMAGES
of America

CUMBERLAND
THE ISLAND CITY

Cumberland is known as the "Island City," as the main downtown section of the community is surrounded by Beaver Dam Lake. Within a 10-mile radius, there are 50 lakes, many filled with record-size fish. Once a major lumbering center, today it is a mixture of industry and recreation. (Author's collection.)

ON THE COVER: The lake crew for the Beaver Dam Lake Lumber Company stands on the company tug boat that moved logs along the lake to the mill. From left to right are Joe Johnston, Bob Brinly, Herb Daniels, George Brown, Harry Luff, Bill Drake, Ed Klink, and Frank Dillon. (Courtesy of Thomas St. Angelo Library.)

IMAGES of America

CUMBERLAND
THE ISLAND CITY

Brent T. Peterson

ARCADIA
PUBLISHING

Copyright © 2019 by Brent T. Peterson
ISBN 978-1-4671-0334-3

Published by Arcadia Publishing
Charleston, South Carolina

Printed in the United States of America

Library of Congress Control Number: 2018965511

For all general information, please contact Arcadia Publishing:
Telephone 843-853-2070
Fax 843-853-0044
E-mail sales@arcadiapublishing.com
For customer service and orders:
Toll-Free 1-888-313-2665

Visit us on the Internet at www.arcadiapublishing.com

This book could not have happened if I was not lucky enough to have parents and grandparents wise enough to purchase a summer lake cabin in the Cumberland community. The summers spent there with family and friends made me, and our entire family, realize what a special place Cumberland is. So, this book is dedicated to my grandparents Frank and Evelyn Peterson and to my parents, Roger and Darlene Peterson. Through all the fun, the cabin on Big Dummy Lake has now seen four generations of the family and will continue to create memories for years to come.

Contents

Acknowledgments 6

Introduction 7

1. Cumberland's Lumber Era 9
2. Cumberland's Business 21
3. Schools, Churches, and Government 43
4. In the Neighborhoods 67
5. Parades, Pageants, and Celebrations 87
6. At the Lake 97
7. Myths and Legends 111

Acknowledgments

The majority of the photographs in this publication have come from the Thomas St. Angelo Library in Cumberland. The library is considered the cultural center of the community, and the director Rob Ankarlo and the library board have been very helpful in gathering the photographs and the information about them. The Local History Room at the library has a great collection and is a community research center like no other in the region.

Mark Fuller was a great asset to the book. He has written a column in the local paper for over 30 years and has knowledge of the community history like no other. John Wickre, a retired employee of the Minnesota Historical Society in St. Paul, has deep roots in Cumberland and knowledge of the railroads that have come through the community.

Connie (Chubb) Bussewitz, her sister Cheryl Ann (Chubb) Cloyd, and brother Bob dove deep into family history and shared some great family photographs. Bob and Carol Sirianni, former owners of the 10-24 Club, were also very helpful in organizing some of the history of the community.

Judy (Christensen) Miller shared family history, including about her dad, who is in the Cumberland Sports Hall of Fame. Charlotte Carlson shared her husband's photograph as police chief. Gary Borman lent photos and stories of his family's past in Cumberland.

I must also acknowledge Louie, of Louie's Finer Meats, for help with the history of his business. The *Cumberland Advocate* has a great archive of local history; thank you for the use of that archive.

Other photographs have come from other organizations such as the Barron County Historical Society located in Cameron, Wisconsin. The historical society is an untapped resource for local history and should be looked at as one of the finest in Wisconsin's northwest. Community individuals who have opened their family history and photographs have also been important to the creation of this publication. Family photographs from 40-plus years of great summers have been used to show the progression of the history of Cumberland and the surrounding area.

Introduction

Cumberland is known for its summer cabins and great fishing. The community is more than that. The area was used by the Ojibwa Nation for hunting and trapping, and when settlers started coming, they used the lakes to help transport the big stands of lumber to mills built on the shores of Beaver Dam Lake.

Many settlers began to pour into the area when the railroads came through. On January 24, 1880, the *Stillwater Messenger* wrote of the new community:

> Cumberland, Wis., the terminus of the North Wisconsin road, is a lively burg. Twelve months ago there was not a frame house in the present village. Now there are five general stores, two blacksmith shops, a shoe shop, a hotel, and numerous boardinghouses. Though in the thick woods it is something of a wheat market, 2,000 bushels having been purchased within three weeks. Three trains leave daily for Hudson, two of them hauling logs, and the other hauling passengers, freight and wood.

The early years saw Cumberland as a rough town; the train through the area was known as "Cumberland, Hayward, Hurley & Hell," which spoke to the lack of law enforcement through those communities. Cumberland quickly hired marshals to enforce the laws and became a peaceful town.

Cumberland's population quadrupled from 1880 to 1890 and continued to grow until the Great Depression of the 1930s. Agriculture replaced the lumber industry as it moved farther west to find more timber. The community had always been known for the area lakes, but it was a resurgence in the 1940s of cabin ownership that brought people to the region.

Cumberland is not a sleepy recreation center, but a town that has been and continues to be a center of industry and a great place to grow up and live. The year-round activities and major businesses that are a part of the community will continue to be a fixture in the region for decades to come. The city might be located on an island, but it is by far not isolated from modern technology.

Today, the residents of Cumberland celebrate their history with the annual Rutabaga Festival and with a large Fourth of July fireworks display. The city of Cumberland has had some ups and downs, but through it all, it remains one of the favorite locations in Wisconsin to live, work, and play.

One

Cumberland's Lumber Era

Northwestern Wisconsin had been a source of timber for logging and lumber companies for several decades before permanent settlement started in the region. Timber cruisers, those who estimate the amount of potential lumber for the company, eventually were able to reach the area where Cumberland was settled.

Once news of the old-growth trees in the area spread, people soon started to populate the area, helped by the railroad. The immense tracts of pine and hardwood timber, which included oak, maple, basswood, and birch, gave rise to some large mills.

In the spring of 1880, a complete steam sawmill was constructed by Mansfield & Lang at what was then North Cumberland. They manufactured their own lumber. From this time, the community began to grow. The firm operated its mill for about two years, until S.G. Cook & Company purchased it. They leased it for a short time, then took it over. The mill burned in August 1884, and N.L. Hunter bought the site and rebuilt the mill. He continued in operation until folding the company in the early 1890s.

In the fall of 1880, Stone & Maxwell constructed another mill. The next year, the mill merged with the Cumberland Lumber Company. On January 1, 1888, the company became a part of the Beaver Dam Lumber Company with John F. Miller as vice president and general manager. In the spring of 1891, this company erected a new band mill and made extensive improvements to the old mill. The mill's capacity after the improvements was 60,000 feet of lumber and 175,000 shingles per day. The company employed 125 men in the mill at that time, and the annual lumber cut was about 12 million board feet.

The mill burned down in 1898 when a lack of rain led to many large fires in the area. The mill was immediately reconstructed, but burned down again and was rebuilt once more. It finally closed in 1911 due to the depletion of timber. Other lumber and logging companies in the area moved west, where the timber had not yet been harvested.

The community survived the end of the lumber era as other manufacturing and businesses that were drawn to the area continued to flourish and thrive. Soon other forms of business would fill the void left by the lumber industry, making sure that Cumberland would have a strong economy and continue to grow and prosper.

In the late fall, logging companies would construct camps for the winter work of cutting down trees, stacking, and sending them to the water's edge. A typical camp included the bunkhouse, cook shanty, offices, and other buildings as needed. This camp near Sand Lake shows the lumberjacks in front of the buildings. Note the cook at far left. (Courtesy of Thomas St. Angelo Library.)

Lumberjack is a general term, as each man had a specific job to do. These two undercutters are notching a tree on a certain side to make it fall in the desired direction. This was done before the sawyers started to saw the tree down. It took real skill to be an undercutter, and those in Northern Wisconsin were among the best in the country. (Photograph by Charles A. Zimmerman, courtesy of Washington County Historical Society.)

Winter work for the Beaver Dam Lumber Company lumberjacks included decking and stacking logs on a sled after they were cut. These sleds would later be pulled to the water's edge. From left to right are Frank Riley, John Wick, and "Poker Jack" Kasabin. (Courtesy of Thomas St. Angelo Library.)

After the trees were felled, they were cut into lengths and loaded onto sleds that were pulled by either horses or oxen. To load these sleds with massive logs took skill and nerve. If the sled of logs was to be larger than most, a pulley system was used as shown here. These men are loading sleds at Gordon, Wisconsin, and the man leaning on the peavey to the right is Alec Bell. (Courtesy of Washington County Historical Society.)

The load of logs was then pulled by two, four, or six horses or oxen on an iced road to the place along the shores of the lake where the logs would be banked for the winter. In the spring, the logs were pushed into the water for transportation to the mill. This is a load of logs for the Beaver Dam Lumber Company around 1890. (Courtesy of Thomas St. Angelo Library.)

Pictured here is the Beaver Dam Lumber Company Camp at Larrigan Lake during the winter of 1896–1897. This load of logs is being pulled by a four-horse team, towing up a long grade known as a "Tow Hill." Another loaded sled is behind the first. (Courtesy of Thomas St. Angelo Library.)

Pictured is the bunkhouse for the Gore & Stinson camp at McKinley. Lumberjacks are sitting on benches while their clothes hang to dry above them. Note the hanging caulked boots, which had spikes in the soles for better grip as they moved from one log to another. (Courtesy of Thomas St. Angelo Library.)

Seen here is the Beaver Dam Lumber Company Larrigan Lake Camp cook shanty. The cook was the most important position in a lumber camp during the winter. Good food served fast made the lumberjacks full and ready for another long afternoon in the woods. The cook was Al Root (right) and his assistant, the "cookie," was Pete Anderson. (Courtesy of Thomas St. Angelo Library.)

Banking the last log of the season by the Gore & Stinson Company of Chippewa Falls, Wisconsin, in March 1906 is pictured here. Over the course of 10 years, the company banked over 200 million log feet, which it drove down the Clam River, into the St. Croix River, and delivered at Nevers Dam. This company operated from two to five camps in the Cumberland area, employing on average 200 men a season. (Courtesy of Thomas St. Angelo Library.)

Lumberjacks from the Sand Lake Camp pull a felled log to a clearing to have it stacked on a sled along with other logs to be brought down to the water's edge. These lumberjacks have the tools of their trade with them, holding peavies, which help roll and move the log. Note at right the camp mascot doing his part to keep up morale. (Courtesy of Thomas St. Angelo Library.)

Channels were created between several lakes to make the transfer of logs from one area to another as easy as possible. At times, these channels needed to be dredged to keep them open. Teams of horses were used to dredge this channel between Kidney and Beaver Dam Lakes around 1905. (Courtesy of Thomas St. Angelo Library.)

If a water channel between lakes was not an option, an above ground transfer was created to move the logs from one lake to another. This shows the Sand Lake end of Beaver Dam Lumber Company's log transfer between Sand and Kidney Lakes. Pictured in the foreground is F.W. Miller, and to the left are Mrs. Miller and their children. (Courtesy of Thomas St. Angelo Library.)

The landing crew is on the lower end of Sand Lake for the Beaver Dam Lumber Company. The shack was a warming house used by the crew between loads. From left to right are Tim Cohan, Dave Davis, John Bisonette, "Poker Jack" Kasabin, and John Wick. (Courtesy of Thomas St. Angelo Library.)

Pictured is the lake crew of the Beaver Dam Lumber Company sometime in 1907–1908. From left to right are Joe "Tony" Johnston, George Brown, A.T. Brinkley, Bob Brinkley, and Ed Klink. (Courtesy of Thomas St. Angelo Library.)

The Beaver Dam Lumber Company tug boat and crew pushed and directed the logs through the lake to the mill. From left to right are Ed Klink, Mike Paulson, Bill Drake, George Brown, and Bob Brinkley. (Courtesy of Thomas St. Angelo Library.)

Logs on Beaver Dam Lake are being readied to go to the mill to be milled into lumber. The logs were held in a pen until it was time to mill them. (Author's collection.)

The first sawmill constructed in Cumberland was by Mansefield and Lange in 1879–1880. The mill was acquired by W.L. Hunter in 1883 and operated by him until about 1892, when it went out of business. (Courtesy of Thomas St. Angelo Library.)

This sawmill was built in 1880 by Maxwell and Stone as the Cumberland Lumber Company. They were joined by Heath and Royce and became the Beaver Lake Lumber Company. Griggs, Foster, and Miller became interested in the firm, and in 1887, became sole owners. The seven renamed the company the Beaver Dam Lumber Company. The mill burned in 1898 but was rebuilt and operated until 1911. (Courtesy of Thomas St. Angelo Library.)

Pictured are the Beaver Dam Lumber Company saw mill boiler room workers around 1907. From left to right are (on the ground) John Nesvold, Pat DeGidio, Axel Dahl, George Johnston, Bert Brinkley, Oscar Helstern, Joe Johnson, Happy Hognes, Cy Lovaas, and two unidentified; (on the ladder) Alfred Johnson; (by the stack) Mike Newman, Henry Holm, Smoky Hognes Dutchy, and unidentified; (on the roof) Jack Lovaas and Bill Woodhouse. (Courtesy of Thomas St. Angelo Library.)

The year 1904 was one of the best and most successful logging seasons in Northern Wisconsin. The lumber camps of the Beaver Dam Lumber Company of Cumberland logged nearly 10 million log feet. The unusually heavy snowfall of the season made the winter one of the best for logging. This photograph shows two sleds of logs coming through Cumberland in front of the Hines residence with the Cumberland Hotel in the background. (Courtesy of Thomas St. Angelo Library.)

The Hunter Saw Mill, above, was one of the largest producing lumber mills in the region. The Beaver Dam Lumber Mill, below, was one of the largest employers in the area until it closed in 1911. (Courtesy of Thomas St. Angelo Library.)

Two

Cumberland's Business

The first pioneer homesteaders came to the island in the mid-1870s. O.A. Ritan and R.H. Clothier, along with Gunder O. Dahlby and his family, came to the island in late 1874. In April 1875, G.C. Hodgkins came and settled in the area. Later in 1875, another family—A.J. Cooke, his wife, their son George, and others arrived—spending their first winter, 1875–1876, in a tent.

A few years later, the first lumber mill was organized and constructed on the shores of Beaver Dam Lake. Soon the lumber industry was in full operation, cutting the trees from the forests in and around the region. This created the need for other businesses to come into the growing community to support the operations of the lumber companies.

During the summer of 1878, David Ingle and his brother opened the first grocery store in a shanty erected by some of the men constructing the railroad in the area. A Mr. Gregg established the first dry goods store.

The railroad did not reach Cumberland until the fall of 1879. The station was constructed immediately afterward, and the track was in working operation in the spring of 1880.

As part of the Beaver Dam Lumber Company, the Company Store was established and was the largest building in Cumberland for many years. It served the company as well as the community by furnishing supplies of all kinds for nearly 100 years. The third floor of the building was known as Miller Hall and was used for plays, lectures, meetings, and other gatherings. It had an estimated seating capacity of 400.

The first bank in the community began in October 1883, known as the State Bank of Cumberland. Another bank was established as a private bank on July 1, 1896, and was known as the Island City State Bank; it was later reorganized and took the name of the Island City Bank.

The Uecke Land Office brought the availability of land in the area to a wider audience. This helped to bring in more settlers and more businesses, which helped after the end of the lumber era. The Box Company, Stella Cheese, and then 3M and Falcon Drill plus the Ardism Company gave the Cumberland community a solid manufacturing base, creating a thriving economy for the area that continues to this day.

A wagon is parked along the wooden sidewalk in the business district of Cumberland around 1910. In the background is the P.H. Peterson Hardware Company, and to the right of that is the Company Store, which opened in 1879 and continued until 1971. (Courtesy of Thomas St. Angelo Library.)

A group of six young people huddle together during the late fall in a wagon pulled by two horses on Second Avenue just in front of H. Woodcock's Home Bakery. (Courtesy of Thomas St. Angelo Library.)

Pictured here is the Cumberland Supply Company, which is next to the Johnson & Ecklie General Merchandise store on Second Avenue. The dirt street has turned to mud after a recent rain. The wagons and horses are being readied for supplies and other items purchased at the two businesses around 1905. (Courtesy of Thomas St. Angelo Library.)

This is Cumberland's main business street, Second Avenue, looking north with several forms of transportation on the street—horses and wagons along with a couple automobiles. Note the electric streetlight above the intersection at center. (Courtesy of Thomas St. Angelo Library.)

Pictured here is a modern barbershop in Cumberland in 1901 with a double sink in the center of the room along with three chairs for patrons. From left to right are George Irwin, Fred Peck, and Herb Daniels. Note the framed print above the middle mirror with Pres. William McKinley in the center decorated with two flags at the top. (Courtesy of Thomas St. Angelo Library.)

The Pioneer Meat Market, located at 1100 Second Avenue, sold more than just meat. Note the Post Bran Flakes sign along with the jar of pickles on the counter. Canned goods line the shelves behind and alongside the counter. The large Miller-Olcott Company calendar is turned to February 1928. John Engesether is at left, and Frank Ecklund is on the right. (Courtesy of Thomas St. Angelo Library.)

Cumberland's iceman Thomas Alfonse stands in his delivery wagon with a pair of ice tongs in his hands. Alfonse, born in Ateleta, Italy, in 1882, came to America in 1885. He married Elizabeth Ranallo in 1907, and in 1917, Alphonse purchased the ice business from George Miller. He leased his ice operation to Ed Timblin and later sold the business to George Ficocello, retiring from the ice business after 30 years. The Pioneer Meat Market is at left across the street. (Courtesy of Bob and Carol Sirianni.)

Pictured here is Tom Alfonse's ice house at the south end of Beaver Dam Lake. It was where the ice was stored, being covered with sawdust for insulation, for the summer trade. Constructed in 1930, it collapsed during a storm around 1960. (Courtesy of Bob & Carol Sirianni.)

Fred Miller, S.H. Waterman, and F.L. Olcott incorporated the Cumberland Telephone Company on February 19, 1898. About 1905, E.V. Benjamin purchased it and conducted it until his death in 1919. L.W. Benjamin succeeded his father as treasurer and general manager. The company had free service with the McKinley Telephone Company and toll connections with the Barron County Telephone Company. J.F. Benjamin became manager after the death of his father, L.W. Benjamin, in 1935. Louis Benjamin is seen at right. (Courtesy of Thomas St. Angelo Library.)

In 1884, there were 24 saloons within the limits of Cumberland. A liquor license in 1885 cost $200. On April 7, 1896, under Mayor W.B. Hopkins, the city voted against the license of saloons. This was the first public action taken in Cumberland in the direction of prohibition, which as a local measure proved only temporary, as the wet and dry forces were at this time about equally balanced, the city going wet again the following year. The records show complaints against certain saloonkeepers for running disorderly houses, and the council investigated all such cases, and occasionally, a license was revoked. The vote on the license question cast on April 6, 1897, was in favor of license by 139 to 134 against. At this time, the cost of a saloon license was $500. The Bentson & Johnson Saloon is pictured here around 1895. (Author's collection.)

A box factory reused the old steam shingle mill along the shores of Beaver Dam Lake. It was located on a three-acre site, which later expanded to six acres. The main type of box made at the factory was for fruit packaging. The factory opened in 1910, and with many additions and an occasional fire, employed members of the Cumberland community until it closed in 1946. (Courtesy of Thomas St. Angelo Library.)

Pictured are box factory employees at the main entrance for the employees. The company continued to employ local residents through its 37 years of operation in Cumberland. (Courtesy of Thomas St. Angelo Library.)

Pictured in the box factory office are the company's secretary Herbert Ewald (right) with his brother Julius. The Ewald brothers were connected with the factory for most of its history. (Courtesy of Thomas St. Angelo Library.)

Box factory workers are pictured outside during the winter. The year 1920 was a busy one with the weather affecting the berry crops and the box business. The bumper crops kept the factory buzzing for the entire season, with both Native Americans and local residents keeping the stream of blueberries coming through Cumberland from the northern parts of Wisconsin. Pictured are, from left to right, Alfred Hanson, Jack Corcoran, and Anton Ecklie. (Courtesy of Thomas St. Angelo Library.)

Pete Sirianni's popcorn wagon is on the dirt streets of Cumberland. Sirianni sold popcorn, peanuts, and other snacks out of the wagon. He later started the 10-24 Club in Cumberland and a bottling company in Rice Lake, Wisconsin. (Courtesy of Bob and Carol Sirianni.)

Henry Hemmingson, manager of the Cumberland Supply Store, sits in his carriage with a team of horses in front of August Wolff's harness shop, on the west side of Second Avenue. (Courtesy of Thomas St. Angelo Library.)

29

The architect of the Hotel Cumberland was L.S. Hicks, of Oshkosh, Wisconsin. The hotel originally cost $15,000. It began construction in August 1890. The grand opening was held on March 10, 1891, and the *Cumberland Advocate* declared it the "most brilliant social event in the history of the county!" The hotel was first leased by J.H. Kahler of Northfield, Minnesota, for five years. The hotel, with its 46 rooms, served residents and visitors alike as one of the foremost hotels in Northern Wisconsin. In 1898, it was taken over by Sarah Thomas, her son Lew, and his wife, Mida. They operated it for many years. After Lew died in the 1940s, Mida continued to run the hotel. The hotel continued to serve the public until it closed in 1961 and was purchased by the Northwestern State Bank. It was demolished in 1965, and a new bank was constructed on the site. This photograph, showing the hotel's distinctive towers and verandas, was taken in 1936. (Courtesy of Thomas St. Angelo Library.)

Tom Mason became editor of the *Cumberland Advocate* in 1900. During his six years as editor, he modernized the paper with the purchase of a monotype casting and setting machine. Mason sold the paper to Judge H.S. Comstock in 1906. (Courtesy of Thomas St. Angelo Library.)

Pictured is the *Cumberland Advocate* press and workforce. Pictured are Sammy Hines, Tom Mason, Ida Johnson, Bob Chamberlain, and Miles Chamberlain. (Courtesy of Thomas St. Angelo Library.)

Albert C. Uecke became mayor of Cumberland in 1906. It had been said that Uecke came to Cumberland penniless, but he soon created a real estate business that drew in people from all over northwestern Wisconsin. (Courtesy of Thomas St. Angelo Library.)

Uecke Opera House and Land Office opened in 1903 at 1190 Foster Street. The cost of the building was $10,000. The land office was on the first floor, and upstairs was a theater that could hold up to 180 patrons, "fitted with up to date staging, scenery and seats," according to the *Cumberland Advocate*. (Courtesy of Thomas St. Angelo Library.)

This theater opened as the Zim Zim Theatre in 1921 at 1345 Second Avenue. In 1935, it underwent a renovation and was renamed the Isle Theatre with a seating capacity of 400. It was still operating in 1951 with a slightly reduced seating capacity of 340. The theater closed in 1997, but reopened in April 2008 totally remodeled and with the addition of Nezzy's bar and grill. The ceiling tiles, chairs, and some of the lighting fixtures are original. The theater closed in October 2012. (Courtesy of Thomas St. Angelo Library.)

The Bank of Cumberland was established as a private institution by J.F. Miller and Jeff Heath in 1883. E.V. Benjamin was the first cashier. Soon, Miller became the sole owner and continued until his death in 1892. In 1903, the bank was incorporated as the State Bank of Cumberland. On August 24, 1931, a brazen robbery of the bank took place. The robbers took about $8,000, but while exiting the town, shots were fired by townspeople as well as the sheriff's department. The robbers fired back and eventually escaped. The money was never recovered but was fully insured. (Courtesy of Thomas St. Angelo Library.)

The 10-24 Club was opened by Pete Sirianni. He had a Dr. Pepper bottling business in nearby Rice Lake and used that company's slogan, "drink Dr. Pepper at 10, 2 & 4 o'clock," to name his Cumberland bar. He later passed the club on to his son Frank. (Courtesy of Bob and Carol Sirianni.)

Will Nyman (left) and Birger Johnson are pictured in the workshop of the Nyman Construction Company. (Courtesy of Connie Chubb Bussewitz.)

These Nyman Construction Company workers build a new foundation for the Congregational Church in Cumberland. Note the unique bricks developed by the company for foundation work. They were thought to give better strength and better insulation to the new foundation. (Courtesy of Connie Chubb Bussewitz.)

Nyman Construction Company was one of the main contractors for barns in the area. Over the course of the history of the company, employees constructed and raised over 300 barns in the region. (Courtesy of Connie Chubb Bussewitz.)

The Company Store was founded in 1879 as part of the Beaver Dam Lumber Company. It served as a commissary, furnishing supplies to the mill and lumber camps then operating in the area. It was a landmark in the community and was damaged by fire three times; the last time in 1971, when it burned so badly that the store never reopened and was torn down. This photograph was taken around 1908. (Courtesy of Thomas St. Angelo Library.)

This image of the Company Store's interior shows the quantity and quality of the store's inventory. Store employees Will Miller (left) and E. Johnson take a quick break between customers. (Courtesy of Thomas St. Angelo Library.)

P.H. Peterson Hardware was located on Second Avenue around 1908. There are wooden sidewalks and hitching posts for the patrons' use. Later, Ben Franklin's was located at this site, followed by Peter & Annie's World Market. (Courtesy of Thomas St. Angelo Library.)

The Coop Oil Association was located at 1060 Elm Street. This 1935 photograph shows, from left to right, Tony Hilton, Ernie Markgren, and Jim Nelson. The site had several other oil and gas stations that were well patronized over the years. (Author's collection.)

In 1878, the first hotel opened in Cumberland known as the Cumberland House, built by Nels Jacobson. Jack Collingwood built the Collingwood House, and soon, the Windsor Hotel was erected at 1140 Second Avenue. Nels Paulson operated it in the 1890s, and it was later run by the Johnson family. This photograph is from around 1900. (Author's collection.)

Nels Paulson, who owned the Windsor Hotel for several years in the 1890s, stands next to one of his prize horses next to the hotel around 1895. (Courtesy of Thomas St. Angelo Library.)

Falcon Drill was founded in May 1962 by a partnership of C.R. "Si" Brock and C. Wm. Mossberg. The company began officially manufacturing high-speed twist drills in October 1962 in a building constructed on Highways 48 and 63. After closing for a short time, it reopened in March 1963. Falcon Drill specialized in small-diameter drills. The company closed in the 1980s. (Author's collection.)

Minnesota Mining & Manufacturing, known as 3M, came to Cumberland in May 1950. It began with only 15 people, including Lou Bohn, plant manager, and Les Mathwig, office manager. The plant was enlarged in 1953 and again in 1961. The main product produced at the plant is sandpaper; however, over the years the plant has evolved and diversified into other business such as floor pads, lapping films, microfinishing films, and superabrasives. (Courtesy of *Cumberland Advocate*.)

Louis Muench Sr. was a meat cutter in Chicago. In 1963, he and his wife, Doris, moved to Cumberland to take a position with the Company Store as the meat department manager. Seven years later, he opened his own meat market, Louie's Finer Meats, on Second Avenue. This c. 1976 photograph is of the original location, now the Cumberland Chamber of Commerce. (Courtesy of Louie Meunch Jr.)

Louie's Finer Meats soon outgrew its original location and moved to a new location on Highway 63 in 1978. Several additions have occurred in the years following the move, including an expanded building for the Louie's Lodge liquor store in 2016. (Courtesy of *Cumberland Advocate*.)

40

Louie Muench Jr. stands next to a new Vorton smoker used for making Louie's award-winning smoked sausages. (Courtesy of Cumberland Advocate.)

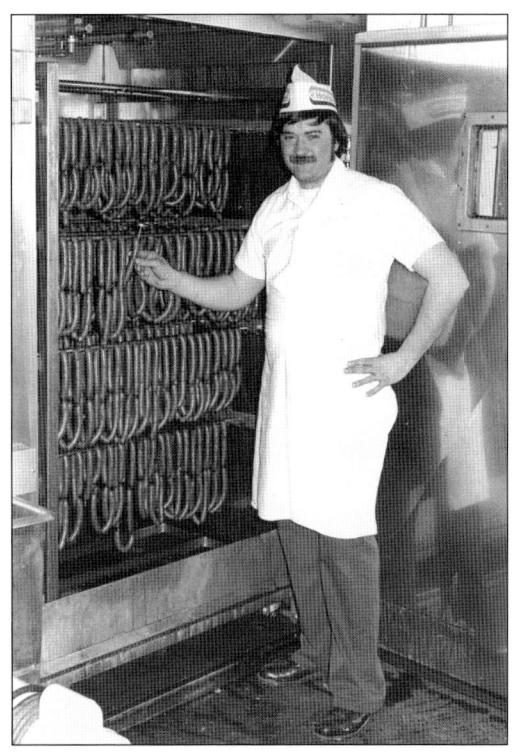

In 1975, Louie's Finer Meats began to participate in competitions. Up through 2018, Louie's has won more than 450 state, national, and international awards. In this 1981 photograph, Jim Muench, left, and Vic Ostrum stand behind the meat counter showing off the award for champion bratwurst in Wisconsin. (Courtesy of Cumberland Advocate.)

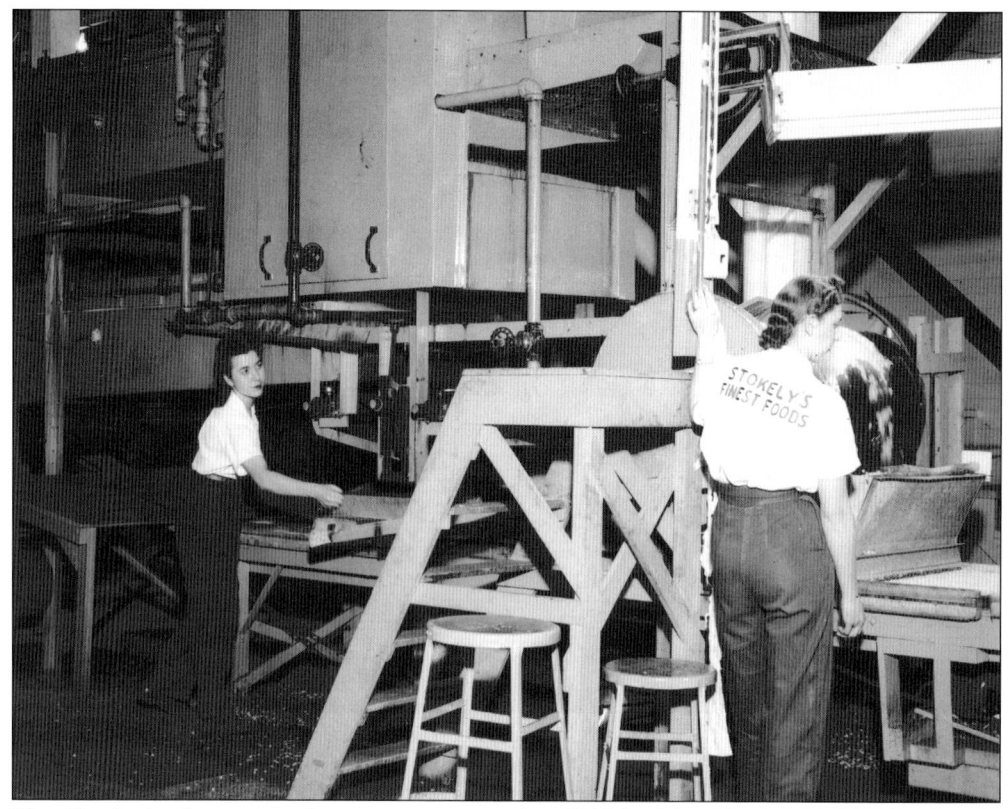

The Cumberland canning plant was constructed in 1912 by John Nimlos stockyards. In 1917, Grafton Johnson combined his three Northern Wisconsin plants to form the Fame Canning Company, which included the Cumberland plant. In 1928, Stokley Brothers & Co. acquired the Fame plant and modernized it. The company canned peas, snap beans, corn, beets, carrots, and other foods. (Author's collection.)

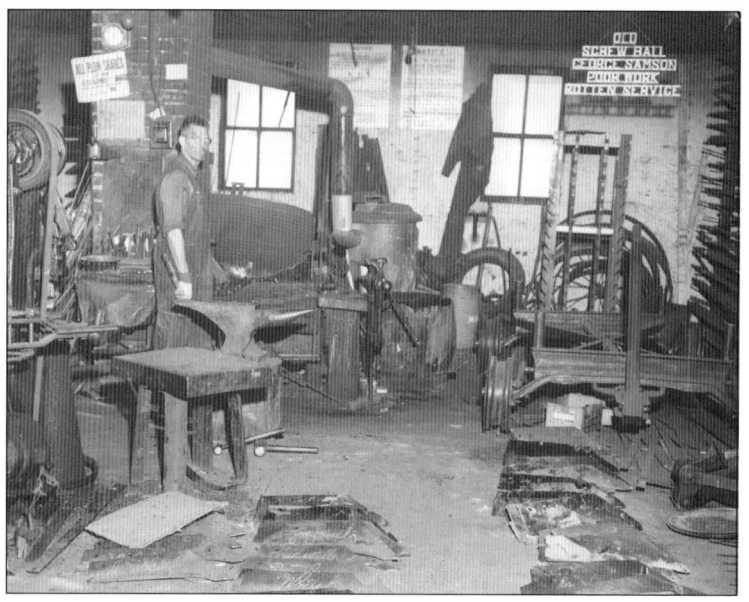

The community blacksmith was an important job. Wagon wheels and farming equipment always needed to be repaired. George Samson was the Cumberland blacksmith for many years. In this photograph, his sense of humor is displayed through the sign at upper right. The repaired plowshares on the floor have customers' names written on them. (Courtesy of Thomas St. Angelo Library.)

Three

Schools, Churches, and Government

In late 1874, Gunder Dahlby made the first claim in what is now Cumberland. He filed a homestead on the north side of the island and constructed a log shanty, a very primitive structure, 12 by 16 feet. O.A. Ritan, who came with him, helped him build it.

A post office was established on April 26, 1876, in Section 10, three miles east of the island with L.L. Gunderson as postmaster. It was called Lakeland. In July 1878, Gunderson moved to the island, bringing the post office with him. The name was changed from Lakeland to Cumberland in 1880. Cumberland was incorporated as a village on November 29, 1881, upon petition of Thomas P. Stone, J.H. Smith, A.D. Fuller, O.A. Ritan, and C.A. Lamoreux. J.F. Fuller was the first village president.

In the spring of 1885, Cumberland was incorporated as a city, with three wards. The first mayor was L.B. Royce. Six years later, the legislature passed an amendment that gave the city of Cumberland practically a new charter; at that time, a fourth ward was created.

The first school was constructed during the winter of 1876–1877, not on the island but just across the bridge. The first teacher was Mrs. Hodgkin, who had begun a school for just her children at first, soon followed by other students. The first district schoolteacher was Ida Schofield, but after several months, Carrie Fay took over and taught the rest of that term, and the next.

The schoolhouse was used for only a few years before a new one was built on the island; it contained four or five rooms and was used as a graded school. As more students attended, the adjacent building of the Masonic Temple was used until a larger school was constructed in 1903. Even with the new building, it became necessary to construct another supplementary building just across the street in 1921. Other larger buildings followed to house the growing population of the community.

In 1927, a new high school was built for Cumberland, and the old school was used as the grade school. The population of Cumberland continued to increase, and another new high school was constructed in 1959.

There were no religious organizations during the early years of Cumberland, but occasionally, a Methodist preacher, or "circuit rider," would stop by the area and conduct services. The first organized church was the Methodist Episcopal, which was organized in October 1882 by Rev. Richard A. Clother.

Lakeland School, seen here around 1892, was located northeast of Cumberland. (Courtesy of Thomas St. Angelo Library.)

Cumberland's earliest high school, on the left, was later used as the Masonic Hall, and the grade school, known as the "white school," on the right, was built in 1881. It was at the early high school that the first Cumberland High School students graduated with a degree in 1891. (Courtesy of Thomas St. Angelo Library.)

The new Cumberland High School was constructed in 1903 at a cost of $35,000. It contained 17 rooms. The *Cumberland Advocate* called it a "handsome and commodious" building and "the most elegant and costly building in this part of the state." (Courtesy of Thomas St. Angelo Library.)

Pictured here is the Cumberland High School class of 1896. From left to right are Emma Johnson, Colista Carsley, Andy Anderson (seated) Sara Ahlgren, and Maybelle Boyden. P.A. Johnson of Cumberland took the photograph. (Courtesy of Thomas St. Angelo Library.)

In 1899, the Cumberland School Board gave principal Lawrence Pease permission to organize a high school football team. The club won three of its four games including a victory over New Richmond. Players pictured here include Lawrence Pease, Percy Morey, Howard Frissell, John Jacobson, Jim Doar, Harvey Helms, Tom Doar, Guy Sampair, Elwood Callahan, Frank Helms, Bert Miller, Oscar Funne, Steve Corbet, and Ralph Hunter. (Courtesy of Thomas St. Angelo Library.)

Pictured here is a Cumberland High School class in May 1909. The enrollment of the schools went from 531 in 1903 in all grades down to 428 in 1915. However, the enrollment in the high school had increased to over 120 in 1915, more than double what it was 10 years earlier. (Courtesy of Barron County Historical Society.)

The Cumberland High School football club practices at right. Football was always a favorite athletic endeavor for students, and games have always been well attended over the years. (Courtesy of Thomas St. Angelo Library.)

Seen here is the Cumberland High School basketball team in 1905. The club is getting ready for an away game, carrying specially marked bags that read "C.H.S.," as well as the player's initials in the lower-right corner and the cheer "Rah, Rah, Rah" at the top. (Courtesy of Thomas St. Angelo Library.)

1912. SENIOR PICNIC.

The high school graduating class always held a senior picnic toward the end of the school year and before graduation. The class of 1912 is pictured on the steps of one of the area's lakeshore cabins so the students could enjoy an afternoon on Beaver Dam Lake. (Courtesy of Thomas St. Angelo Library.)

On the steps of the Cumberland Hotel, several members of the Cumberland High School football club sit for a photograph. Note the hats being worn by the members of the club and the person in the door looking on. (Courtesy of Thomas St. Angelo Library.)

Cumberland's new high school building is at right, along with the old 1903 school. The old high school became the grade school after the new building was completed in 1935. (Courtesy of Thomas St. Angelo Library.)

The new Cumberland High School, constructed next to the old school in 1935, was used until 1959 as a high school. (Courtesy of Thomas St. Angelo Library.)

High school students are outside the Cumberland High School during a break in the day. Note the old high school at left. (Courtesy of *Cumberland Advocate*.)

A high school tradition, prom, takes place at the Cumberland High School in 1938. The tradition of a formal dance continues at the high school today. (Courtesy of *Cumberland Advocate*.)

Pictured here are the Cumberland High School Glee Clubs of 1937–1938. Above is the girls club, and below is the boys club. The two clubs would have several events a year, plus one in which the two performed together. (Both, courtesy of *Cumberland Advocate*.)

Cumberland High School baseball has had a long history. From 1929 to mid-season of 1932, the club did not lose a game. They had a 26-game winning streak until Colfax defeated Cumberland in May 1932. Baseball was dropped from the high school sports for a few years during the Depression and World War II but was reinstated in 1951. The team pictured here is made up of, from left to right, (first row) Fred Wolff, Elwood Callahan, and Leo Wachter; (second row) Bert Hines, F. Carsley, Harvey Helms, and Oscar Funne; (third row) Lawrence Pease, George Hopkins, Ernie Kellerman, Rick Stone, and Mr. Coleman. (Courtesy of Thomas St. Angelo Library.)

Three deliverymen in a wagon from the Company Store stop in front of the Cumberland High School. Many students would work in the area businesses after classes or in the morning to make extra money. (Courtesy of Barron County Historical Society.)

Pictured here are the Cumberland High School cheerleaders around 1937–1938. (Courtesy of *Cumberland Advocate*.)

This is the Cumberland High School golf team of 1937–1938. The team held its home matches at the Cumberland Golf Course, just east of town. (Courtesy of *Cumberland Advocate*.)

John Plichta was a teacher and coach at Cumberland High School from 1928 to 1930. He was born in West Allis, Wisconsin, and attended Ripon College. He taught in Vermont and came to Cumberland in the fall of 1928. He was an avid ski jumper and left a legacy of ski jumping in Cumberland, making a record jump on February 24, 1929, of 61 feet. Plichta was also an avid photographer and took many early 16mm films of Cumberland High School and the community. He left Cumberland in 1930, returning to his hometown of West Allis and teaching history as well as coaching. He later married, but his wife died in childbirth, leaving him and his son; they moved back in with Plichta's parents. Plichta is pictured here. (Courtesy of Thomas St. Angelo Library.)

This is the Cumberland Junior Band during the 1937–1938 school year. Many of these musicians would continue to play in the band during their senior year. (Courtesy of *Cumberland Advocate*.)

The Cumberland Marching Band practices marching on Second Avenue. In the background are the Company Store and the Tofness Chiropractic Clinic. (Courtesy of Thomas St. Angelo Library.)

Cumberland High School band director John P. Anderson graduated from Stevens Point Central College. He was in charge of the band for 37 years from the 1950s to the 1980s. (Courtesy of *Cumberland Advocate*.)

The marching band plays on the field during Cumberland High School's homecoming game halftime. The game between Cumberland and Hayward was on October 5, 1973, with Cumberland winning 18-15. This halftime show included former coaches of Cumberland High School along with past homecoming kings and queens. (Courtesy of Thomas St. Angelo Library.)

Helen "Missy" Adams, a kindergarten teacher, was named National Teacher of the Year for 1961. Pres. John F. Kennedy, who established the award, said, "Miss Adams will stand as an example to future teachers and an inspiration to all Americans for dedication and warmth, and her struggle against high odds to become a teacher." (Courtesy of *Cumberland Advocate*.)

Pictured at the presentation of the National Teacher of the Year award are, from left to right, Walter Gannon, Cumberland grade school principal; Helen "Missy" Adams; Fred Moser, high school principal; and Matt Lofy, English teacher. Lofy was later killed in a car crash in March 1965 while traveling to watch the Cumberland High School basketball team play in the state tournament in Madison. (Courtesy of *Cumberland Advocate*.)

Cheerleaders have been a part of Cumberland athletics for many years. These three are from the junior high in 1969. From left to right are Barb Dawkins, Carol Cifaldi, and Kris Erickson. (Courtesy of *Cumberland Advocate*.)

In 1958, the voters in the Cumberland School District approved plans for a new high school building at the cost of $850,000. It was to be located on a new 52-acre site just southwest of town. The new building was occupied in 1959 with a student capacity of 475. In 1965, a junior high school wing was added to the building. (Courtesy of Thomas St. Angelo Library.)

Pictured is the confirmation class of the Augustana Lutheran Church around 1980. (Courtesy of *Cumberland Advocate*.)

Augustana Lutheran Church was constructed in 1907 when some members split from the First Lutheran congregation. There were several improvements to the church building over the years including removing the steeple, rebuilding part of the tower, installing carpeting, and more. (Courtesy of Thomas St. Angelo Library.)

Swedish Lutheran and Norwegian Lutheran Churches are seen here side by side in Cumberland around 1910. (Courtesy of Thomas St. Angelo Library.)

The First Lutheran congregation was organized in 1876 by Norwegian immigrants under pastor C.J. Helsem. In 1886, First Lutheran united with Lutherans of Swedish descent in building a church. It was completed in 1887 on a site donated by O.A. Ritan. The two congregations shared the building, holding separate services in their own languages, until the Swedish congregation could afford their own church. Rev. A.J. Logeland, pictured here with his family, was pastor of First Lutheran from 1891 to 1906. (Courtesy of Thomas St. Angelo Library.)

The Catholic community in Cumberland first had traveling priests come in for services, which were held at the Company Store. In 1883, a church building began construction under direction of Father DeParadis, and it was finished in 1884. On May 7, 1944, the building burned. A new building was constructed and dedicated on May 4, 1948, with Bishop Meyer officiating. (Courtesy of Thomas St. Angelo Library.)

The Methodist church was constructed in 1882 and dedicated in 1883. Originally, it had a membership of 193, but as other Methodist churches opened nearby, the membership dropped to 86. In 1920, after several important meetings were held at the church, membership climbed to 200 people. The Ladies Aid of the church became well known for their quilt making, and one of their quilts was presented to Grace Coolidge, wife of Pres. Calvin Coolidge. (Author's collection.)

The Cumberland Post Office moved depending on who was the postmaster. It was started in 1876, and the community has had mail service ever since. This c. 1900 photograph shows the post office when it was located at 1404 Second Avenue. (Author's collection.)

The post office interior is shown here with Bert Pease at left and George Johnson at right. Note the folded US flag on the table. (Courtesy of Thomas St. Angelo Library.)

The post office employees seen here are, from left to right, (first row) Selma Johnson and Margaret Pease; (second row) George Johnson and Bert Pease, who was the postmaster at Cumberland from 1907 to 1915. (Courtesy of Thomas St. Angelo Library.)

Pictured is the Cumberland Post Office around 1910. The post office had been located in many different buildings over the years. (Courtesy of Thomas St. Angelo Library.)

Pictured here is the post office staff; from left to right are (first row) Louis Tappon, John Oren, Cecil Stowe, and Andrew Stoll; (second row) Rose Mann, Bert Pease, and Margaret (Norton) Pease. (Courtesy of Thomas St. Angelo Library.)

Pictured here is the post office interior. Note the mail boxes at left and the sorting table in the foreground. (Courtesy of Thomas St. Angelo Library.)

The Cumberland Fire Department began in 1883 with A.H. Kellermann as chief. That year, the community purchased a secondhand hand-operated pumper. "It was a man killer," Kellermann remembered later. The early department, made up of all volunteers, faded away, and in the early part of the 20th century Cumberland did not have a fire department at all. This photograph shows a Cumberland firefighter dousing the flames of a local fire. (Courtesy of *Cumberland Advocate*.)

The Cumberland Fire Department reorganized into a modern and professional department on December 2, 1912. August Wolff led the drive for a new fire department. This photograph shows three Cumberland firefighters holding a hose while fighting a fire. (Courtesy of *Cumberland Advocate*.)

Early Cumberland was a lawless point on the map, but Cumberland soon took steps to secure law and order. A marshal was hired, Patrick Hurley, a former Stillwater, Minnesota, police officer who soon set about cleaning up the town. The well-known "slugfest" between the marshal and Nels Paulsen is still talked about today. Police Chief Bruce Carlson became chief in 1986 and retired in 2001. (Courtesy of Charlotte Carlson.)

Four

IN THE NEIGHBORHOODS

Cumberland, like most other communities, had strong neighborhoods. These areas were created based on ethnic backgrounds to begin, with areas for Swedes, Norwegians, Italians, and others who carved out their own identities.

Although there were different nationalities that came to Cumberland, the Italian community is the best known. The first Italian pioneers were predominantly from the district of Abbruzzi or the province of Aquila of Cantalupo, and the cities of Campobasso and Boiano. Others came from Ateleta and the town of St. Polo Matese. From the extreme southern part of Italy, in the province of Catanzaro, came the Siriannis, Chiodos, and Caliguires. The first Italian settlers in Cumberland came in the late 1870s and early 1880s, and some of their names still fill the pages of the Cumberland phone book. By 1895, much of the Italian settlement was over.

The place for all people to come is the Cumberland Library. At a special meeting of the city council in April 1898, the city council rooms were chosen as the location for a reading room and library. On June 10, 1904, the library moved to the old high school building, with Mrs. G.E. Carr as the librarian. The *Cumberland Advocate* editor Tom Mason corresponded with philanthropist Andrew Carnegie, and on January 26, 1905, the *Advocate* announced the arrival of a letter stating that Carnegie would be happy to give $10,000 toward a library.

Cumberland mayor W.E. Hines and library president W.B. Hopkins laid the cornerstone in September 1905. The *Advocate* exclaimed in the March 15, 1906, issue "Cumberland's handsome and commodious library building will be open to the public, Saturday March 17th." The building continued to be a mainstay in Cumberland. In May 1989, library director Margaret Palmer expressed a need for library expansion, and with individual donations along with some foundation money, the expansion of the library was complete in 2009, and the library was named after local statesman Thomas St. Angelo.

The neighborhoods were also a place to go to scout meetings, play baseball, listen to music, and even go bowling on a weekend night. Catching a train to Spooner or Hayward was a family outing, while staying at home or working in the yard might be another thing to do.

The core of the Cumberland community is not its ethnic background or the amount of money people make; it is the belief that their neighborhoods, police, firefighters, and others are working together to make it the best place to live.

This photograph shows the Miller and Hines homes and the Cumberland Hotel. The Miller house was constructed in 1882 at a cost of $5,000. The original structure consisted of six rooms on the main floor, included a sewing room, a formal parlor and dining room, a powder room, and a kitchen. The second floor had seven bedrooms and a bath. The top floor was used as a game room. Originally, the home was quite simple, but over the years, more architectural amenities were added along with its signature tower. (Courtesy of Thomas St. Angelo Library.)

The first Cumberland Hospital, established by Dr. G.A. Grinde, was located in this home on Grove Street. An operating room was on the second floor, and the house was converted into a 10-bed hospital in 1916. After the first year of business, there were 137 operations performed at the hospital. The house was later extensively remodeled; a six-crib nursery was added and x-ray equipment installed, and the number of beds in the hospital more than doubled. (Courtesy of Thomas St. Angelo Library.)

The Italian Society of Cumberland was organized in 1896. It was called the Societa di S. Antonio Abate, and at one time, there were more than 100 members. The Italian Hall, also known as Columbia Hall, was built in 1907. The society held frequent meetings there and as many as 150–200 people would show up for events. In the 1950s, there were 20 members in the group when it decided to close. (Courtesy of Thomas St. Angelo Library.)

The William and Minnie Nyman residence across from the Augustana Church on South Second Avenue is seen here. This home was constructed with the unique bricks the Nyman Construction Company used for better insulation and support. It still stands as a testament to their quality of workmanship and ingenuity. (Courtesy of Connie Chubb Bussewitz.)

This is the Dr. Grinde house after the expansion to a 22-bed hospital. This house was used as the hospital for Cumberland for many years and was later sold as a private residence. (Courtesy of Thomas St. Angelo Library.)

Men are on a stoop in Cumberland on Second Avenue. Note the wooden boardwalks for sidewalks and the Company Store in the background. From left to right are (first row) Bill Swaner, Johnny Stoll, Ed Brand, George Ellefson, Jim Martell, Ed Klink, Will Lewis, and Josh Tuttle; (second row) Albert Uecke and William Jeffery. (Courtesy of Thomas St. Angelo Library.)

This photograph was taken in front of the Cumberland Hotel in 1908. From left to right are Charles Piske, Sadie Hocom, Grace Mason, George Stahl, Dr. C.E. Foote, Tom Mason, and Walt Hocom. (Courtesy of Thomas St. Angelo Library.)

John Nyman was photographed in a wagon with his sister Nettie in front of the August Wolff building. The Wolff building was constructed in 1902. August Wolff served two non-consecutive terms as Cumberland mayor, 1900–1901 and 1920–1922. (Courtesy of Connie Chubb Bussewitz.)

After the Beaver Dam Lumber Company folded in 1912, Frank Olcott and Will Miller created a partnership in the Miller-Olcott Lumber Company on First Avenue. Olcott resigned from the company in 1928, and Miller died in 1933. The lumber company was sold to the Andersen Corporation in Bayport, Minnesota. It has since become the Lampert Lumber Company. Pictured are Miller-Olcott lumber company workers around 1930. (Courtesy of Thomas St. Angelo Library.)

S.W. Hines came to Cumberland in 1879 and went to work for Griggs, Rhode & Miller and the original owners of the Company Store. He later became a partner in O.A. Ritan & Company, then purchased the firm and established S.W. Hines Mercantile Company. Hines later purchased the Company Store and operated both businesses for a number of years. He was twice elected mayor of Cumberland. (Courtesy of Thomas St. Angelo Library.)

The Hines family lived in a beautiful brick home between the Miller house and the Cumberland Hotel. The house was later used for the offices of the Wickre Agency real estate company. This photograph shows the Hines family around 1925. (Courtesy of Thomas St. Angelo Library.)

The Chris Christensen family is pictured here; from left to right are (first row) Chris, Glen, Cecil, and Henrietta; (second row) Blanche, Mabel, Harold, Hazel, and Charles; (third row) Herbert, Myrtle, and Walter. Chris and Henrietta were married in 1895 in Iowa. They came to Cumberland in 1920 and their sons Glen and Charles were principal owners of Christensen Electric. Cecil was a member of the 1932 Cumberland High School basketball team and was elected to the Cumberland Sports Hall of Fame. (Courtesy of Judy Christensen Miller.)

Nels Paulson was born on August 26, 1847, in Norway, the son of Jergen and Ann Paulson. He came to the United States with his family in 1867 and soon found employment in lumbering for various companies. In 1881, he purchased 120 acres of land eight miles south of Cumberland. On July 22, 1882, he married Caroline Songu. In 1884, he traded part of his land for the Windsor Hotel, which he ran for six years. This photograph shows Nels at left, with his wife Caroline at right, and their two oldest children Tilda (left) and Anna. (Courtesy of Thomas St. Angelo Library.)

Dr. Charles E. Foote is believed to have been the first dentist in Cumberland when he started his practice in 1882. He remained in Cumberland for 27 years. He is pictured here with his wife, Mae. (Courtesy of Thomas St. Angelo Library.)

John Nyman's girls and their aunt Lena (left) are dressed up for Easter in the yard of the family home. Note the Easter lily plant at left. (Courtesy of Connie Chubb Bussewitz.)

Pictured here is the O.A. Ritan family. Ritan, with Gunder Dahlby, made the first permanent settlement on the island that is now Cumberland. He was a successful businessman and built up a large mercantile business. In 1884, he constructed the large brick building that became known in later years as the S.W. Hines Mercantile Company. He was one of the original incorporators of the Village of Cumberland in 1881. (Courtesy of Thomas St. Angelo Library.)

The Cumberland Brass Band was organized in the early part of the 20th century. They played at special events in the community and the surrounding area including dances, celebrations, and family gatherings. Cumberland has a long tradition of community bands and orchestras. (Author's collection.)

The Cumberland Boys Band was invited to go to Superior, Wisconsin, around 1928 to perform for a special event that included Pres. Calvin Coolidge. After the event, President Coolidge stood with the band for this photograph. (Courtesy of Thomas St. Angelo Library.)

The Cumberland Country Club was founded about 1925, operated by the Cumberland Land & Improvement Company. Tom Vardon, the club professional at the White Bear Yacht Club in White Bear Lake, Minnesota, laid out the course. Vardon also designed the golf courses in Amery, Spooner, and many others in Minnesota and Wisconsin. He was also the brother of the famous English golfer Harry Vardon. The parent company operated the course until 1942, when it ceased operations during World War II. It reopened with private ownership, which lasted until the city took over in 1968. The old clubhouse, constructed around 1930, had a large screen porch around it as shown above around 1930; a new clubhouse was opened in 1969. The photograph below shows golfers on Father's Day. (Above, courtesy of Cumberland Golf Course; below, courtesy of *Cumberland Advocate*.)

Above are the Cumberland Boy Scouts of Troop No. 24. From left to right are (first row) Jimmy Statton, Chuck Toftness, Dennis Crawford, and Denny Schneider; (second row) Bob Wasilensky, unidentified, Bob Toftness; (third row) Gordon Toftness (leader), Jim Toftness, Robbie Lund, Larry Alberg, unidentified, and Dave DeGidio. (Courtesy of Thomas St. Angelo Library.)

Pictured here are the Eagle Scouts of Troop No. 24; from left to right are (first row) Mark Nelson, Gordon Toftness (leader), and Steve King; (second row) Sam Donatelle, Tom Toftness, and Brian O'Hern. Gordon Toftness was scoutmaster for 25 years and was awarded the Silver Beaver & Silver Antelope awards. Fritz Hines was the first Eagle Scout of the troop in 1932, and there have been 78 more since then up to 2018. (Courtesy of *Cumberland Advocate*.)

The pinewood derby, a racing event for miniature cars, is run by Cub Scouts. Here, a Cub Scout pack in Cumberland is holding their annual races. (Courtesy of *Cumberland Advocate*.)

Four lanes for bowling were installed at the 10-24 Club in the late 1930s. Originally, there were pinsetters who would set the pins and take away those that were knocked down. An automatic pinsetter was installed in the 1950s. (Courtesy of Bob and Carol Sirianni.)

The ball field in Cumberland was located near the 3M manufacturing plant. Several baseball greats played on this field, including Satchel Paige, Tommie Aaron, and Bob Uecker. (Courtesy of Thomas St. Angelo Library.)

The softball club sponsored by the 10-24 Club is seen here around 1947. From left to right are (first row) bat boy Bob Sirianni; (second row) Gerry Gargaro, Don Nesvold, Clarence Capra, Bud Nesvold, and Hand Hendricks; (third row) Lee Dosch, Carl Stender, Len Mayer, Bill Hanson, Tony Gonzales, Earl Thompson, and Frank Sirianni. (Courtesy of Bob and Carol Sirianni.)

Thomas St. Angelo was born in Cumberland on January 13, 1889, the son of Genaro and Angela St. Angelo. He graduated from Cumberland High School and completed a law course with LaSalle Extension of the University of Chicago. He was on the Palmer School Board from 1920 to 1927, with Federal Farm Credit Administration 1933–1957, and served two terms in the Wisconsin Assembly from 1960 to 1964. He died on June 16, 1967 at Cumberland Memorial Hospital. (Courtesy of Thomas St. Angelo Library.)

Cumberland organized a library in 1898. The city council appropriated money annually and provided rooms to be used as a public library. In 1904, the city requested funds from Andrew Carnegie for a new library. Funds were granted, and the library building, designed by Minnesota architect C.H. Patsche, was constructed at 1305 Second Avenue. It was completed in 1906. The building was listed in the National Register of Historic Places on June 25, 1992. In 2009, it was expanded and renamed the Thomas St. Angelo Public Library. (Courtesy of Thomas St. Angelo Library.)

This view of the Cumberland business district looks north on Second Avenue; the library is on the left. (Courtesy of Thomas St. Angelo Library.)

Pictured here is the children's area in the library around 1965. Librarian Katherine Robinson is second from right. (Courtesy of Thomas St. Angelo Library.)

The longtime commander of Anderson-Thomson American Legion Post No. 98 in Cumberland was Jim Schweiger, pictured here. (Courtesy of *Cumberland Advocate*.)

The Cumberland American Legion post was granted a charter on October 10, 1919, and became Post No. 98. It was named in honor of Homer L. Anderson, who was the first Cumberland serviceman to lose his life in World War I. Anderson was a 1914 graduate of Cumberland High School and died when his ship was torpedoed and sunk in the Irish Sea. On February 26, 1946, the post voted to add the name of Alex Thomson to the official name, as Thomson was the first local serviceman to be killed in World War II. (Courtesy of Thomas St. Angelo Library.)

Seen here is the winter toboggan slide in Cumberland Park. Note the ice rink in the foreground and the small kitchen at right that served hot food and beverages to the crowds that gathered there. (Courtesy of Thomas St. Angelo Library.)

The toboggan slide in Cumberland became well known in the area as one of the longest slides. It was constructed by the city and was used in the park for many years. There were two parallel tracks—one with scaffolding at the top for longer and faster runs. (Author's collection.)

The snow did not deter the people of Cumberland, no matter how deep it got. Here, "Swede" Nelson is not bothered by the deep snow as he digs his car out of a snow bank. (Courtesy of Thomas St. Angelo Library.)

The tracks pictured here in the winter led to the Standard Oil gas station. The station featured Red Crown gasoline. Located on the corner of Second Avenue and Elm Street, the gas station remained there for many years. (Author's collection.)

Five

Parades, Pageants, and Celebrations

All across the nation, there are celebrations of Independence Day on July 4. In every corner of the country, there are fireworks, parades, and picnics, and the community of Cumberland is no exception.

Fourth of July is a special time of the year in Cumberland. The day's fireworks have been a mainstay in the town for generations. Even in the earliest days of Cumberland, this holiday was looked forward to and celebrated like no other.

In the fall, the Cumberland High School homecoming football game is also a highlight of the season. There was always much excitement around the team, playing for many years at Moser Field, and whether they were going to win against one of many nearby opponents for bragging rights for another year. Along with the game itself, there was the pageant for the homecoming queen and king and their court, a long parade down Second Avenue with themed floats and other dignitaries, and many after-game events.

In the early days of Cumberland, one of the root vegetables grown and stored in the area was the rutabaga. Known as the "Swedish turnip," the rutabaga was honored in 1932 when the community started an annual summer celebration. It became known as Rutabaga Days and continues to this day.

The event had everything, from fireworks to street dances as well as festival royalty. The parade in 1952 had about 40 units. That year, Frank Sirianni was the parade chair, and Albert Skinner and Bob Harkness were the parade marshals. There were 25 floats and about a dozen or more musical groups that walked the route and entertained the crowds of onlookers. Just 10 years later, the parade chair in 1962, Pat Doonan, had 75 units in the parade with many coming from as far away as Eau Claire, Wisconsin.

In the winter, ice-fishing contests have been popular. One other winter event popular with the people of Cumberland is the annual Dunk the Clunk contest. A car is placed on the frozen Beaver Dam Lake, and people guess when it will break through the ice in the spring. The winning guesser gets a prize—not the car, but something just as nice.

Many festivals and celebrations over the years have been interrupted by bad weather. Just as the post office claimed that neither snow, rain, ice, or the gloom of darkness would stop the mail from being delivered, nothing stopped the community of Cumberland from celebrating whenever they could.

This is a bird's-eye view of Cumberland looking southwest. The main street, Second Avenue, runs diagonally through the image. From right to left are the Methodist church, Cumberland Hotel, the Hines residence, the Miller residence, and the Company Store. It is this street that parades follow. (Courtesy of Thomas St. Angelo Library.)

Sabatino Donatelle is pictured in his delivery wagon. Donatelle started a grocery store on the south end of town in 1888. He made house calls with this horse and wagon loaded with groceries. His son—and later, his grandson—took over the business. The photograph shows the wagon and Sabatino ready for the 1913 Fourth of July parade in Cumberland. (Courtesy of Thomas St. Angelo Library.)

The 1916 Fourth of July parade included many unique floats including this entry for George Irwin's barbershop. Sim Brand drives while Lee Hines shaves Dewey Jacobson at front. Zean Douglas shaves Peter Knutson at the rear of the wagon, and Mart Wick is lighting the big make-believe firecracker. Gordon Finley is with the baby pig sucking a milk bottle. (Courtesy of Thomas St. Angelo Library.)

Homecoming at Cumberland High School has always been a big celebration. The homecoming parade is always an event that is looked forward to. Here, the float of the "Beaver Machine" is getting ready to move into the lineup. (Courtesy of *Cumberland Advocate*.)

All decked out for the parade is the P.L. Ferguson truck with Standard Oil and Red Crown gasoline advertised. It must have been for a Fourth of July event, considering all the flags attached to the truck. (Courtesy of Thomas St. Angelo Library.)

The Rutabaga Days parade is one of the largest in Wisconsin's northwest. In 1949, the Mayer-Rose Lumber Company had this exquisite float featuring a large pheasant on display. The float is in front of the Isle Theater. (Courtesy of Thomas St. Angelo Library.)

The Women's Broom Brigade is ready to clean up the town. This photograph is from an unknown parade or community event, but the women may have been supporting women's suffrage. (Courtesy of Thomas St. Angelo Library.)

The Falcon Drill Team consisted of employees of the Falcon Drill Company of Cumberland. The team marched in parades in Cumberland and around the area for several years. (Courtesy of Bob and Carol Sirianni.)

Pictured here is Jim and Klem's Hobo Band. This band played for many years at the Cumberland Rutabaga Days festival. From left to right are Jim ?, Klem ?, two unidentified, and A.B. Clock from Cumberland on the bass drum. (Courtesy of Thomas St. Angelo Library.)

Well-known Cumberland resident Jack Richie marches in the 1950 Rutabaga Days parade dressed as one of the local baseball stars. Note the oversized bat and shoes. (Courtesy of Bob and Carol Sirianni.)

Second Avenue in Cumberland is all decorated and lit up for the holiday season in 1940. The Cumberland Library is at left. (Courtesy of Thomas St. Angelo Library.)

Pictured here is the Cumberland fire truck leading the Rutabaga Days parade, with the Cumberland High School cheerleaders on top. The Rutabaga Festival started in 1932 and continues to be the community's major celebration. (Courtesy of *Cumberland Advocate*.)

Pictured here are the Rutabaga Days royalty in the parade. Royalty have been chosen at the community festival for most of its history. (Courtesy of *Cumberland Advocate*.)

This image of the 1973 Rutabaga Days parade shows the floats and parade attendees lining Second Avenue and cheering the participants on. On the left is the Gustafson Ice Cream shop, and the larger building is the old Uecke Opera House. (Courtesy of Thomas St. Angelo Library.)

Bowling tournaments were a big event during celebrations in Cumberland. Here, the Gutter Belles of Stokely Canning are about to bowl against the Mining Beaverettes of the 3M Corporation in Cumberland. (Author's collection.)

The first annual hot pepper eating contest sponsored by the Spot Bar at Rutabaga Days was held in the parking lot behind the bar. Much excitement was created with this new contest, and a lot of milk was drunk afterward. (Courtesy of *Cumberland Advocate*.)

Pitching horseshoes has been a very popular event throughout the years of Rutabaga Days. The contest, sometimes held at the city park, has drawn contestants and spectators from miles around. These two champions show off their trophies. (Courtesy of *Cumberland Advocate*.)

Six

AT THE LAKE

Cumberland, being surrounded by Beaver Dam Lake, has long been a destination for people to spend the weekend or longer at one of the many resorts that dot the bays on the lake. Summer residents flock to the community, making the population swell and the atmosphere sparkle with excitement.

Some families had cottages on the lake in the early years, but it was W.G. and A.H. Miller who started a business renting cottages out to visitors. Their cottages were built in an area known as Miller's Sunset Point and had a living room, two bedrooms, kitchen, pantry, and a fully screened front porch. A good well for water was available. The renters were taken out to the cabins from Cumberland on a boat, and the rental fees were $10 a week.

As demand grew, more cottages were constructed. The Huntingtons built about 20 cabins on the north shore of Beaver Dam Lake, which were usually rented out to clergymen. Other places around the lake included Ernie Miller's cottages, the Miller farm at the head of the lake, and the Hines and Miller cottages in the same area.

Other nearby lakes had resorts as well. In the 1940s, Art Fisher created the Granite Lake Resort on Granite Lake. He sold the resort to Sig Gullikson, who later sold the business to Eric Just in 1952. At that time, the resort was renamed the Five O'Clock Club.

Interest in settling in Cumberland increased. As more people came to rent a summer cabin, many decided it could be a good place to settle. Albert Uecke began reaching out to people to purchase tracts of land in the early 20th century. Later, Lloyd Wickre became the real estate person known for "selling Cumberland."

John D. Olson was a major landowner and sold much real estate in the area. In 1921, William Talbot joined Olsen in his land and lumbering company. Talbot made his first real estate transaction in 1921. In 1941, he started a retail business for International Harvester farm machinery and other light industrial equipment. The company continued through three generations.

Within 10 miles of Cumberland there are nearly 50 lakes with residences, resorts, or campgrounds. The community continues to be the center of the place where people come to visit and create memories for both young and old.

Being at the lake meant a lot of social gatherings and get-togethers. Above, at the head of Beaver Dam Lake, high school students take time to camp out. The photograph below shows Lydia Mill and Bert Miller holding a pole. Note the hammock above and the tents at right below. (Both, courtesy of Thomas St. Angelo Library.)

The Cumberland municipal bathing beach on Beaver Dam Lake has been a longtime destination for locals and visitors alike. This image shows the diving dock in the water but also the two dressing rooms—one for boys and the other for girls. (Courtesy of Bob and Carol Sirianni.)

Showing the Cumberland beach from another angle, this image features the diving dock and a couple rowboats that were available for fishing. (Courtesy of *Cumberland Advocate*.)

Pictured here is the W.G. Curtis landing on Beaver Dam Lake. The September 10, 1896, *Cumberland Advocate* mentioned, "W.G. Curtis has launched his steamboat the 'Edna' and has built a boathouse to house it on Beaver Dam." (Author's collection.)

W.C. Curtis's steamer *Edna*, named for his granddaughter, was 34 feet long and could hold up to 50 people. It was later acquired by Zean L. Douglas and rechristened the *Mary Ann*. (Courtesy of Thomas St. Angelo Library.)

Two girls pose among birch trees on the west side of Beaver Dam Lake, looking east. (Courtesy of Thomas St. Angelo Library.)

Anna and Sarah Peterson fish off the dock at the family summer residence on Big Dummy Lake just north of Cumberland on Highway B. Their dog, Murphy, watches every move of the bobbers to see if a fish is on the line. (Author's collection.)

Fishing is one of the biggest pastimes at the lake. Crappies, bass, walleye, or northern pike are the trophies of vacationers and locals as well. *Sports & Recreation* magazine of March–April 1963 reported on the bountiful fishing in and around Cumberland: "Norwegian Bay of Beaver Dam Lake was especially good," according to the article. Earl Meyer of Cumberland had six walleyes that weighted nine and ten pounds and others in the seven- and eight-pound range. John Lillie caught two seven pounders and was pictured in the article with one of the fish he caught. The article also mentions Bear Lake, just to the north of Cumberland, and Vermillion Lake, just south of town, and how they were yielding some large fish as well. In this c. 1925 photograph, three men show off their catch from Beaver Dam Lake. (Courtesy of Thomas St. Angelo Library.)

Bert Miller caught fish on July 30, 1904, from one of the Miller cabins on Beaver Dam Lake. The size of northern pike like these is what has drawn people for many years to fish Beaver Dam Lake. (Courtesy of Thomas St. Angelo Library.)

Local anglers "Pistol Berg" (left) and George Stoll are pictured with the day's catch. Stoll is holding his trusty fishing pole, and in the back, a canoe is leaning up against their automobile, in which they surely took to their favorite fishing hole. (Courtesy of Thomas St. Angelo Library.)

Marshall Skinner stands next to the day's hunting success. Two deer are hanging in the tree at left. (Courtesy of *Cumberland Advocate*.)

Hunting was and still is an annual event in the area. Many local residents and visitors take time to go deer hunting every season. From left to right, Martin Wick, Tony Gidio, and Howard Steinburg had great success during the 1948 season. (Courtesy of *Cumberland Advocate*.)

Victorian boating is pictured here on Beaver Dam Lake. The proper clothing was necessary for a woman to be out on a boat with a gentleman. These three seem to be having a wonderful afternoon on June 20, 1897. (Courtesy of Thomas St. Angelo Library.)

A steamer is landing at the dock for Camp Dixon on Beaver Dam Lake. The steamers for the resorts would take guests around the lake and even land near town for dinner or an afternoon of shopping. (Courtesy of Library of Congress.)

Miller cottage No. 3 sits high on a hill overlooking Beaver Dam Lake. The Miller family had several cabins that it rented during the summers on the lake. (Author's collection.)

Frank Olcott had a large stretch of land beyond the Cumberland Golf Course and a pleasure boat that he christened the *Idlewild*. He sold his property in 1928 to Emil Kokes of Illinois, who was a friend of Chicago mayor Anton Cermac. The property was later sold to Frank Roman, who converted the area into a lodge, naming it after the boat *Idlewild*. This photograph shows the view from the Idlewild Lodge. (Courtesy of Thomas St. Angelo Library.)

The municipal beach in Cumberland is very popular. The diving dock has been a fixture at the beach for decades, and there is a lifeguard floating nearby in a boat in case of emergency. The beach continues to be a source of relaxation and a place to cool off during the heat of summer. (Courtesy of *Cumberland Advocate*.)

For many years, the municipal beach has been the place for swimming lessons. Many locals remember taking their first strokes as a swimmer in the waters of Beaver Dam Lake at the beach. (Courtesy of *Cumberland Advocate*.)

The days of steamboat excursions have long since passed and have been replaced with more modern modes of transportation, including the pontoon boat. Here, a family or group of friends travel near the shores of Beaver Dam Lake, feeling the breeze in their faces. (Courtesy of *Cumberland Advocate*.)

The Peterson family purchased this cabin and property on the north shore of Big Dummy Lake in 1974 and made it their summer residence. The lake was said to have been named for a local Native American chief who was deaf and "dumb" and eventually drowned in the lake. (Courtesy of Jeff Peterson.)

The most important item for ice fishing is an auger to drill a hole in the ice to fish through. In Cumberland, a local company named Ardisam, incorporated in 1960, produces powered ice and earth augers along with adaptors that covert chain saws into augers. The company began in two leased buildings in downtown, but in 1971 moved to its new quarters on Highway 63. Three people began the company, but Richard Ruppel bought out the other two to become the sole owner. "Our augers make both fun and work a lot easier," said Ruppel. Ardisam augers are used all over the United States and throughout the world. (Courtesy of *Cumberland Advocate*.)

Sitting on a bucket with a line in the hole, anglers wait for a bite as they participate in the annual ice-fishing contest on Beaver Dam Lake. (Courtesy of *Cumberland Advocate*.)

To some, sitting outside on buckets is not the proper way to fish during the winter. Many winter anglers build icehouses, which they drag onto the lake and sit inside. The ice is thick enough for them to drive right up to their icehouse and park their truck. (Courtesy of *Cumberland Advocate*.)

Seven
MYTHS AND LEGENDS

In Cumberland, like in most places, there are stories of legendary residents or events that have been handed down from one generation to another. Most of these stories have a thread of truth, while much of the rest is made up. Whichever it might be, these are the myths and legends that the community hangs their history on.

The circus has long been a favorite form of entertainment. The Ringling Brothers have history in Wisconsin, and the famous family actually lived in Rice Lake for a time, just a few miles west of Cumberland. The Yankee Robinson Circus stopped in Cumberland in 1911 and helped make one of those stories that will never be forgotten. It was in June that year that the lead elephant of the circus, known as Tom Tom, drowned in Beaver Dam Lake.

Julius Alfonse, a well-known young man in the community and a standout athlete at Cumberland High School, went on to play football with the University of Minnesota in 1936 and won the national championship. The Cleveland Rams drafted him in 1937.

Sports were a great escape in the community and some high profile athletes of the past have played in Cumberland including the great Negro League and Baseball Hall of Fame pitcher Satchel Page, Tommie Aaron when he was with the Eau Claire Braves, and Bob Uecker when he was with the same club in 1957. Hall of Fame pitcher Burleigh Grimes grew up in nearby Clear Lake, Wisconsin, and threw out the first ball of the Island City Tournament in 1982.

John and Ben Peterson, brothers who graduated from Cumberland High School and lived in Comstock, were wrestlers who won Olympic medals in both the 1972 and 1976 Olympics.

There are as many myths and legends of Cumberland as there have been residents and visitors to the "Island City." These memories are created at high school football games, on the lake with a parent or sibling, or just having a bite to eat in one of the local restaurants.

The Native Americans called the Cumberland area home for many years before the settlers started to move in. In the 1850s, a band of natives left their original area on Lake Court O'Reilles in Sawyer County on a trip that took them south. They continued moving down the Cedar River to Rice Lake, then west to the Yellow River, where they camped before moving on to a lake they called "Way-ko-ne-ma-daw-wang-gog," meaning the "Lake of the Beavers"—now Beaver Dam Lake. One of the men, known as Little Pipe, took up a homestead on Sand Lake. He had a scar on his upper lip and was known to early settlers as Cut Lip. He is shown in a canoe with a friend around 1895. Little Pipe drowned in Beaver Dam Lake when his canoe capsized in 1897. His wife was with him but was able to make it to shore. (Courtesy of Thomas St. Angelo Library.)

Bill Borrman was a master cheesemaker in Cumberland and owned a farm in the area for many years. He married Emma Westerlund in 1922. Their first child, Margaret, was born on April 21, 1923, and was a premature birth; the local doctors told them to prepare for the baby's death. Not satisfied with the diagnosis, Bill went to the local Indian reservation to see if one of the medicine men could come back to the farm and help with treating the baby. Emma was not in favor of the plan, and was terrified when two Indians came into her kitchen. The medicine man asked where "the white papoose was" and went upstairs and performed some medical rituals, then turned to the mother and said, "papoose will live." They were right; Margaret lived until she accidentally drowned when she was 25 years old. (Both, courtesy of Gary Borrman.)

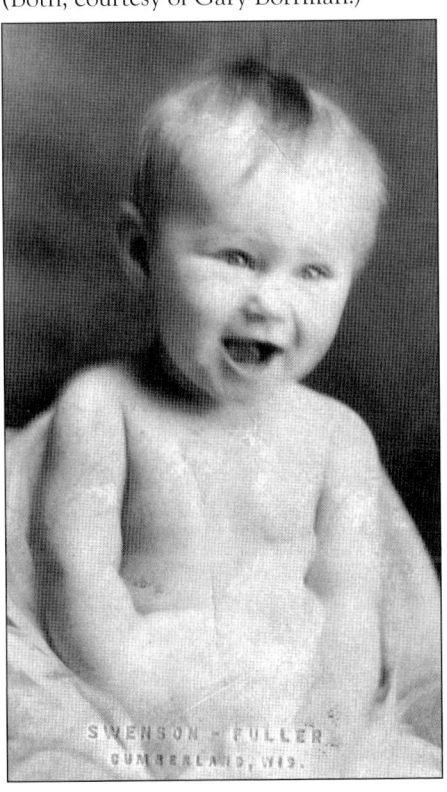

Cumberland, Wis. – 1897

Robert M. LaFollette Sr. barnstormed across Wisconsin during his campaigns for governor and US Senate; he is shown here in Cumberland in 1897. He ran for Wisconsin governor in 1896 and again in 1898, losing both times. In 1900, he won the seat and went on to be successful in the next two elections. He was elected to the US Senate by the state assembly in 1905, where he served until his death in 1925. LaFollette Sr. was an outspoken leader of the progressive side of the Republican Party. After a successful re-election, he helped to start the National Progressive Republican League, an organization devoted to passing progressive laws such as primary elections and the direct election of US senators. After his death, his son Robert LaFollette Jr. took his seat in the Senate. His younger son Phillip served several years as Wisconsin governor. (Courtesy of Wisconsin Historical Society.)

Sr. Mary Madeleva was born in Cumberland on May 24, 1887, the daughter of Mr. and Mrs. August Wolff. She attended Cumberland High School, then the University of Wisconsin in 1905–1906, followed by St. Mary's College at Notre Dame. She graduated in 1909 and entered the Congregation of the Sisters of the Holy Cross. (Courtesy of Thomas St. Angelo Library.)

Eva Wolff (right) is seen with classmate Anna McMalin with a basketball during their high school days in Cumberland. Wolff graduated in 1904 and became Sr. Mary Madelva. She began a career as an author in 1923 with her first book, *Knight Errant and Other Poems*. She authored another 17 books after that and was recognized internationally as a poet and scholar of distinction. She died in July 1964. (Courtesy of Thomas St. Angelo Library.)

Crowds line the shore of Beaver Dam Lake to watch boat races. These races were one of the draws for summer visitors as well as local residents. (Courtesy of Thomas St. Angelo Library.)

The largest craft to ever ply the waters of Beaver Dam Lake was the steamer *Island City*. Used as both a passenger and excursion boat from the city to the head of the lake, it operated from the late 1890s to 1905, when it was sold for scrap for $60. (Courtesy of Thomas St. Angelo Library.)

Depot Scene, Cumberland, Wis.

The North Wisconsin Railway Company was organized in November 1871. It was authorized to construct a railroad from the St. Croix River at Hudson to the west end of Lake Superior at Bayfield. The first segment was built to New Richmond, and by 1874, the line had reached Clayton. A three-year delay caused the railroad to not reach Cumberland until 1878. The following year, the railroad completed another 26 miles. The railroad was vital for Cumberland. A passenger train ran between Cumberland and the Twin Cities every day the first year. In 1919, there were three passenger trains daily, increasing to eight a day in the 1920s. Following World War II ridership decreased, and passenger service was discontinued in the 1950s. (Courtesy of Thomas St. Angelo Library.)

Plaster "doctor" John Till became widely known in the region for his plaster cures of all kinds. Special trains came to see the doctor in his offices, located on a farm just outside of Almena, Wisconsin. Till was later forced to leave the county only to come back and die in Kiel, Wisconsin, in 1947. (Author's collection.)

The local governments tried to close down Till several times, arresting him for practicing medicine without a license. Since Till never claimed to be a doctor, these charges were dropped. There were also lawsuits against him ranging from $100 to over $100,000. It was not long before the local newspapers that once promoted Till started to turn against him. "It would seem if this man," said the *Stillwater Messenger* of August 3, 1907, "is an imposer and a swindler that the authorities of St. Croix County should make an effort to protect the public against him." (Author's collection.)

Fred Moser came to Cumberland in the fall of 1925 as a history and economics teacher and as the athletic coach for the high school. During his first three years, the students, as well as school officials, had learned to trust him for his ability to assume great responsibility. Moser graduated from River Falls High School at the age of 16. He attended River Falls Normal School and graduated from there in 1915. His first teaching position was at Red Granite High School at Red Granite, Wisconsin. He enlisted in the army during World War I and served for over a year. After the war, he taught at Minong, Wisconsin, and at Elmwood, Wisconsin. He received his PhD from the University of Wisconsin. Moser was named principal of Cumberland High School in 1929. Seen here is a c. 1935 photograph of Fred and Katherine Moser. (Courtesy of Thomas St. Angelo Library.)

In the fall of 1888, Bert Pease, editor of the *Cumberland Advocate*, decided to get married. His father, Dr. W.C. Pease, was a physician for the railroad and managed to get some passes for his son and new wife on the train for their honeymoon. He was gone for several days, and when he returned, he found that he had been replaced as editor of the paper. Pease soon became the editor of the *Glidden Pioneer*, but had to move the paper to Ashland, Wisconsin, on a moment's notice, which was done. He then moved to Spooner, Wisconsin, and started the *Washburn County Register*. He later moved that paper to Shell Lake, the county seat. Pease then returned to Cumberland and started the *Free Press* in 1902. Three years later, the paper was moved to Amery, Wisconsin, becoming the *Amery Free Press*. Pease again returned to Cumberland and served as postmaster from 1907 to 1915. Pictured here from left to right are Bertha Pease, Bert Pease, and Ada Uecke. (Courtesy of Thomas St. Angelo Library.)

John and Ben Peterson, sons of Mr. and Mrs. Paul Peterson of Comstock, were both graduates of Cumberland High School. The two brothers wrestled their way to the 1972 Olympics in Munich, Germany, where Ben won a gold medal, and John won a silver. This image shows John Peterson wrestling for Cumberland High School. (Courtesy of *Cumberland Advocate*.)

John Peterson gets ready to go to the 1976 Olympics in Montreal, Canada, again with his brother Ben. In 1976, John won a gold medal, and Ben won silver. (Courtesy of *Cumberland Advocate*.)

Friends enjoy watermelon in the studio of photographer P.A. Johnson of Cumberland. Pictured are Clara Miller, Maud Waterman, Catherine Hollister, Ed Tuckey, Milla Ritan, Bess Hamilton, Jean Evans, Lydia Millen, Guy Maxwell, and Jay Hamilton. (Courtesy of Thomas St. Angelo Library.)

The Yankee Robinson Circus came into town on June 24, 1911. The wagons and performers created a stir in the community. Men, women, and children came down to the depot to see the circus unload and march up to its camping site. The big elephants drew a large crowd, and the leading elephant, Tom Tom, led the pack. Tom Tom stopped at the side of the lake, dipped his trunk in for a drink, and then, for some reason, walked forward into the lake. The handlers thought the elephant was taking a bath but soon saw the big animal struggle and drown. It took four horses to pull him out of the lake. The large tusks were cut off, and Tom Tom was buried nearby. (Courtesy of Thomas St. Angelo Library.)

A sled is being pulled by oxen in front of August Wolff's harness store. The driver is unidentified, but standing in the sled are, from left to right, Roy Smith, "Spec" Lewis, Will Lewis, Frank Dillon, and Karl "Smokey" Hogness. (Courtesy of Thomas St. Angelo Library.)

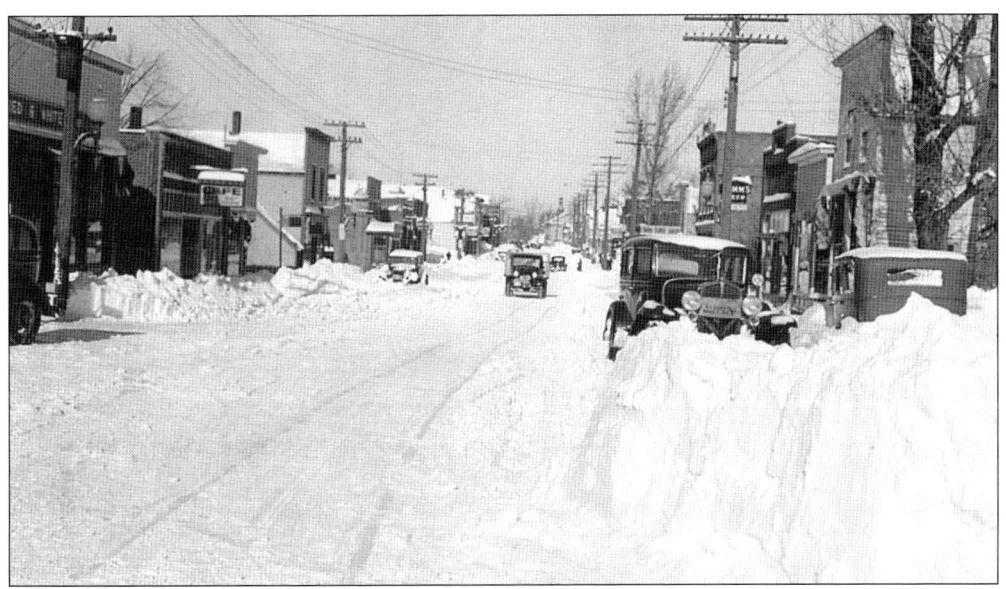

This image of Cumberland's Second Avenue in the winter shows automobile tracks through the snow with some autos buried in the snow. Note the streetlights on the telephone poles. (Courtesy of Thomas St. Angelo Library.)

Sled races on the streets of Cumberland were common during the winter. Many young men tried their hand and wanted to become a legend by besting their opponents. (Courtesy of Thomas St. Angelo Library.)

The *Ralph* tugboat was used by the W.L. Hunter lumber company on Beaver Dam Lake to tow log rafts to the Hunter Mill in the 1880s and early 1890s. W.L. Hunter is at right with his foot up on the bow, and Ralph Hunter is to the left of him with the dog. The two young boys at bottom are unidentified. The boat was later sold and shipped to Washburn, Wisconsin. (Courtesy of Thomas St. Angelo Library.)

Pictured here are dentist Dr. John Cronwell (left) and Dr. Hedback in their office in Cumberland around 1910. Dr. Cronwell was born July 10, 1886, at Perley, Wisconsin, and was a graduate of Cumberland High School in 1907. (Courtesy of Barron County Historical Society.)

The rectory of the Catholic church was constructed along with the new church building. It later became a bed-and-breakfast. (Courtesy of Thomas St. Angelo Library.)

Beaver Dam Lake is seen with low water in the 1930s during the region-wide drought. Many lakes and rivers had their lowest recorded levels during this time. Mr. and Mrs. E.L. Miller are shown in this photograph. (Courtesy of Thomas St. Angelo Library.)

The 1903 high school was used as a grade school after the new school was constructed in 1935, but eventually after another new school was built. The old school, in which so many memories were created, was demolished in 1971. (Courtesy of Thomas St. Angelo Library.)

The Wickre Agency was a mainstay in Cumberland for many years. Lloyd Wickre, a 1937 graduate of Cumberland High School and a 1941 graduate of the University of Minnesota, joined C.H. Kilbourn's agency in 1949, and the firm became the Kilbourn-Wickre Agency. When Kilbourn retired in 1956, Wickre became the sole owner and changed the name to Wickre Agency. In 1968, he moved the agency to the Hines home next to the Tower House. (Courtesy of *Cumberland Advocate*.)

Sam and Clara Momchilovich bought the gas station from Don Tait in 1941. A grocery store and living quarters were added and began business that same year. Sam also sold parts for machine repairs in the old building until another facility was added in the mid-1940s. He later sold Oldsmobiles at this location. This photograph was taken around 1945. (Courtesy of Bob and Carol Sirianni.)

Discover Thousands of Local History Books
Featuring Millions of Vintage Images

Arcadia Publishing, the leading local history publisher in the United States, is committed to making history accessible and meaningful through publishing books that celebrate and preserve the heritage of America's people and places.

Find more books like this at
www.arcadiapublishing.com

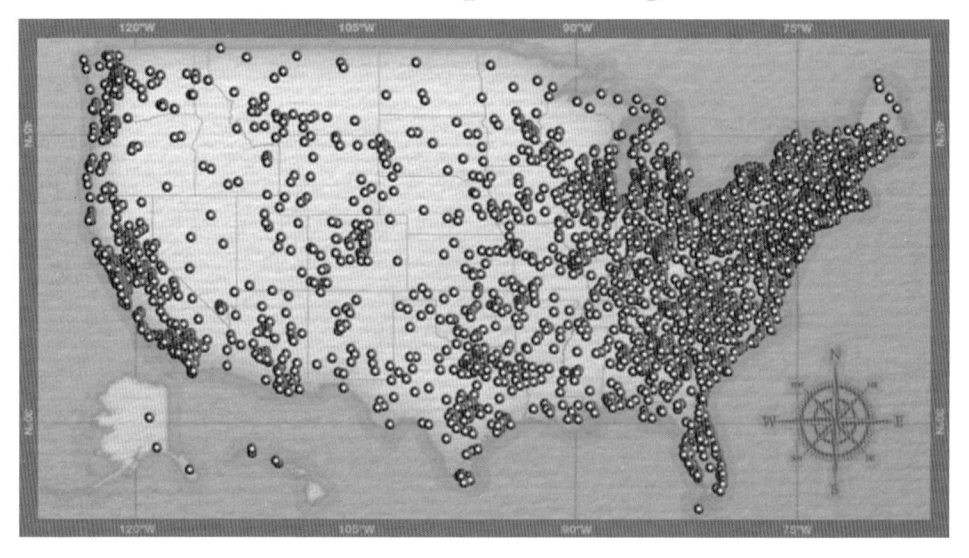

Search for your hometown history, your old stomping grounds, and even your favorite sports team.

Consistent with our mission to preserve history on a local level, this book was printed in South Carolina on American-made paper and manufactured entirely in the United States. Products carrying the accredited Forest Stewardship Council (FSC) label are printed on 100 percent FSC-certified paper.

MADE IN THE USA